They Found
Order in Nature

by Erica Stux

They Found
Order in Nature

by Erica Stux

ISBN-13:978 1530758876

I wish to thank my best friend,

Susan Hanna, for typing my manuscript,

and also for her contributions

to my essays on Galileo

and Richard Feynman.

Books by Erica Stux

Landlady

Eight Who Made a Difference: Pioneer Women in the Arts

Writing for Freedom: The Story of Lydia Maria Child

Sequins and Sorrow

Enrico Fermi, Trailblazer in Nuclear Physics

The Achievers: Great Women in the Biological Sciences

The Wonder of Wings

Incredible Insects

Reflections on Abraham Lincoln and Other Poems

Expressions of Nature through Photography and Words

Naturally Inspired: Poems of the Great Outdoors

Permutations of the Humble Coffee Bean: Poems of Daily Tasks and Diversions

Table of Contents

Pages

Table of Contents

COPERNICUS

A period in the history of western civilization dating roughly from the 1400s to the mid-1600s is known as the Renaissance, a term meaning rebirth. During this period many people in Europe studied the writings of philosophers of ancient Greece and Rome. Ancient Greek thinkers believed that the earth was the center of the universe, and that the sun, moon, and all the stars revolved around the earth. This idea was slowly replaced by the observation that the motion of certain stars and planets did not make sense according to this earth-centered view of the universe. A different model of the universe was needed.

One of those who offered a new model that placed the sun at the center of the universe was a young man named Nicolaus Copernicus. His ideas changed the way people thought about the world. That was the beginning of modern astronomy.

Nicolaus Copernicus was born on February 19, 1473, to a couple living in the town of Torun in Poland, an important port on the Vistula River. His father was a merchant dealing in copper, which made the family wealthy. When Nicolaus was ten years old, his father died, and his mother's brother began helping the family; this man, Nicolaus'uncle, was a Catholic bishop who lived in a huge castle. He arranged for Nicolaus and his older brother

Andrew to attend the University of Krakow, one of the oldest universities in Europe. There, Nicolaus studied math, geometry, and astrology; astrology deals with the supposed effect that the stars and planets have on people's lives. For this part of his education Nicolaus learned to observe and record the motions of the planets. From Krakow, Nicolaus and Andrew went to Italy and enrolled in the University of Bologna, where Nicolaus was supposed to study law, but instead he learned Greek and more astronomy. He also studied church law, probably because in the year 1500 the Catholic Church was celebrating coincidentally a five-hundred-year anniversary. He was offered a job in the Church, but decided to continue his studies, this time concentrating on medicine. After finishing his studies in Italy, he returned to Poland.

Nicolaus' uncle, the bishop, had his nephew serve him both as personal physician and secretary. This proved to be a comfortable life. But in his spare time, after nightfall, he studied the heavens. His observations showed that the earth-centered theory of the universe could not explain his data; he had to find a different model. His calculations led him to a model that explained many motions of stars; his model also demoted earth from the center of the universe to the role of a planet like Mars or Jupiter, a planet that rotated, causing day and night, while the sun stood still. This model explained the order and spacing of the planets: Mercury and Venus closer to the sun than Earth; Mars, Jupiter, and Saturn farther away. (The telescope had not yet been invented, so scientists did not know of planets Uranus, Neptune, or Pluto.)

After his uncle died in March 1512, Nicolaus Copernicus took on the duties of a church official called a canon, which gave him many responsibilities. One project he worked on was the standardization of money in Poland. Silver coins would be melted down by individuals, the silver mixed with cheaper metals, and new coins made from the mix. Copernicus warned people about the cheapened coins, but many of his suggestions were ignored. He was kept very busy as a physician, especially when epidemics swept through the country in 1519.

The pet project of Copernicus was his sun-centered model of the solar system. He wrote an account of his theory, but it was never published in his lifetime. For it to be accepted in scientific circles, he would need mathematical details. He finished writing his theory around 1530, but kept revising his work. Meantime, he distributed hand-written unrevised copies among his friends. He was afraid people would make fun of his ideas, and he was right about that.

In May 1539, a young man who called himself Rheticus convinced Copernicus to offer his theory to the world. He found a publisher in the German city of Nuremberg; the publisher gave the work a Latin title meaning "On the Revolution of the Heavenly Spheres", and added a preface that called Copernicus' theory of a heliocentric system a hypothesis, which meant the author himself (Copernicus) did not really believe that his theory was true. It was published in March 1543. Shortly thereafter Copernicus suffered a stroke and lapsed into a coma. He was barely alive when a copy of his book arrived from the publisher, and then he died.

In the early 1600s, the Catholic Church published a list of books that Catholics were not allowed to read. Copernicus' book was included in the list in the year 1616. Over two hundred years passed before the Church took the book off its list, in 1820. By that time most Catholics had accepted the model of our sun-centered solar system.

A Digression by Susan Hanna

Is the Earth or the Sun the center of the solar system?

Up until Ptolemy and Copernicus observed the stars and planets in the night sky, tracing their motion and appearance, scientists thought the earth stood still and everything including the sun revolved around it. This was known as the geocentric theory; the prefix geo- is the Latin root for land or earth.

This scientific theory complemented man's growing sense of his own importance on the planet earth: his work in a bustling city, the development of the local economy, trade with other individuals from neighboring countries whose language, customs and clothing were unique to their own culture. This period in the history of self-aggrandizement lasting from 1450 to 1600 was called the Renaissance (ren uh zance); and artists and sculptors like Leonardo da Vinci, Raphael, and Michelangelo who lived then in Italy celebrated the importance of man in their art.

Everyone thought since the earth was man's domain, by logical deduction it must be that the earth is the center of the universe. With the Church's teaching and approval, the geocentric theory reigned for centuries. To challenge this theory was blasphemy.

Science and religion have always been great adversaries. Science bases its ideas on reason, evidence, experiments and provable mathematical facts; and religion places its faith in miracles and spiritual mysteries which cannot be properly described. Science and religion for centuries have always clashed. This conflict was never more serious than during the Renaissance.

No scientist dared speak out against Church teachings, although they agreed with Copernicus' theory of 1530 that the sun and not the earth was the center of our universe. This new sun-centered model of the universe was called the heliocentric theory, the prefix helios- being the Greek root for sun. Nicolaus Copernicus died in 1539 before the Catholic Church could condemn him as an heretic.

TYCHO BRAHE

Tycho Brahe was a Danish astronomer who designed instruments in order to obtain positions of the planets. (The telescope was not invented until 1608.) Born into a noble family in 1546, Tycho was raised by an uncle. When he was thirteen, astronomers of that time predicted a solar eclipse on a date in August of 1560, which greatly impressed Tycho.

He attended classes in law and the humanities at several European universities, but his major interest remained astronomy. After finding a miscalculation in the data of Copernicus, he decided to improve what was known regarding positions of the planets and stars. For this he built a private observatory in 1571. His discovery and report of a supernova (the explosion of an entire star) in 1573 made him famous.

He studied and recorded the position of planets at many points in their orbits, and carried out the first research on comets. His careful measurement of the sun's apparent movement led him to report that the year according to the Julian calendar then in use was actually too long, and accordingly ten days were dropped in 1582 and in succeeding years.

Tycho's final years were spent as Imperial Mathematician under Emperor Rudolf II in the city of

Prague. His data was later used by his assistant Johannes Kepler. Tycho died in 1601.

JOHANNES KEPLER

Johannes Kepler was among a group considered modern scientists because they used math in their scientific work. Kepler believed that the movements of heavenly bodies could be expressed by math. His work led to the discovery of the three laws of planetary motion, which became an important part of modern astronomy.

Kepler was born in the German area called Wurttemberg on December 27, 1571, as the oldest of seven children, of which only three lived beyond their childhood. His intelligence was evident from an early age. He attended a school where instruction was in the German language, but soon he was transferred to a school where the study of Latin prepared students for university studies, and in Kepler's case for a career as a clergyman in the Lutheran Church.

His mother introduced him to astronomy by taking him outside to view a comet, and a lunar eclipse in 1580. He was fiercely competitive in his studies. In his teens he had a number of physical ailments, which he recorded by writing about himself in the third person. In September 1589 he began studying at the University of Tubingen, where he increasingly concentrated on math and astronomy. He was introduced to the idea of a sun-centered universe, which explained planetary motion. The university turned Kepler into a gifted mathematician. He was ready to enter the clergy, but was offered a position to teach math at a small seminary in present-day Austria. Accepting it would mean giving up his plan to enter the clergy, but he believed it possible to return to it at some later date.

Kepler had so few math students that he was asked to teach writing, and the classical literature, as well. He felt himself inadequate as a teacher, and wondered if he had made the right decision not to enter the clergy. He turned again to astronomy.

In the spring of 1597 Kepler at age 25 published his first book <u>Cosmological Mystery – The Secret of the Universe</u>. In it he wrote that the sun provided the force that moved the planets along their orbits. He had to buy and distribute 200 copies himself. Responses from scientists varied from enthusiasm to disagreement.

When hostilities broke out between Catholics and Protestants, the local ruler Archduke Ferdinand ordered that Protestant clergy and teachers had to convert to Catholicism, or leave the area immediately. Kepler was among those who left, but then was allowed to return. He had concluded that the Sun provides a force on all the planets. But how powerful was it, and how did it vary over distance?

Kepler was introduced to Tycho Brahe, a Danish astronomer, when Brahe was made official mathematician at the court of Emperor Rudolph II in the city of Prague. Kepler needed all of Brahe's observations, and Brahe need Kepler's genius to make sense of his observations. Brahe died in November 1601, and Kepler was made the new imperial mathematician. He was to complete the astronomical tables that Brahe had started.

Kepler's book <u>Optical Part of Astronomy</u> discusses the nature of light and the process of vision, as well as distances from Earth to the Moon, Sun, and planets. This book helped to confirm his hold on the position of imperial mathematician.

Kepler's work on mapping Earth's orbit and Moon's orbit showed that planets do not move at uniform speeds, but will sweep out over equal areas in equal time intervals. This became known as Kepler's Second Law of Planetary Motion. Kepler's First Law states that planets travel in elliptical, not circular, orbits. His Third Law connects the distances of the planets from the Sun to their orbital periods.

Kepler's book <u>New Astronomy</u> was published in 1608; few of his contemporaries realized its importance, but it allowed scientists to accept the heliocentric system that Kepler described, and it made Kepler the most famous astronomer in Europe at that time.

GALILEO GALILEI

Galileo is sometimes called the first modern scientist because he used experiments and measurements to support any theory. He believed math should be used to understand and describe physical phenomena.

He was born in Pisa, Italy, on February 15, 1564. In 1574 his family moved to the Italian city of Florence, and soon thereafter Galileo began attending classes to learn Greek, Latin, and logic. At age seventeen he enrolled in the University of Pisa as a medical student. A family friend taught him some math, which caused Galileo to neglect his medical classes in order to pursue math studies. In this he was so successful that at age twenty-five he was appointed chairman of the math department at the University of Pisa. Three years later, an influential member of the Italian nobility helped Galileo get a similar but high-paying appointment at the University of Padua. Besides his math lectures, he also taught astronomy.

Galileo's years in Padua, from 1592 to 1610, were the happiest of his life. To help meet expenses, he rented rooms in his house to students, tutored his boarders, and built and sold scientific instruments, especially a military compass.

In 1609 Galileo learned of a new instrument made in Holland: a tube that magnified distant objects for a viewer. Galileo, realizing its military value, made a better one and demonstrated his telescope to the ruler of Venice and the city's senate. The senators were impressed; they rewarded Galileo with a lifetime appointment as state mathematician, with a generous salary.

Galileo pointed his telescope at the Moon and saw, to his surprise, rather than a smooth sphere, that the Moon had mountains and valleys. His telescope also showed him that the planet Jupiter had four moons circling around it and that the planet Venus passed through distinct phases, thus proving that it orbits the Sun, just like the Moon's changing shape shows that it orbits the Earth.

Galileo began to write a pamphlet he called "The Starry Messenger". Although written in Latin, his 550 copies sold out rapidly and became a sensation in Europe.

To combat scientists like Galileo who had mathematics and modern instruments like the telescope at their disposal, the Catholic Church created the Inquisition. This was a special panel of men, usually friars, who had the authority to prosecute, torture and imprison all who presented ideas contrary to the accepted teachings of the Catholic religion. Galileo considered himself a devout Christian who would never question the Catholic Church's teachings.

When Galileo was ready to go to Rome in 1611 to present his work to the pope and other church officials, he was advised, in order to avoid conflict with the Church, to present his discoveries with the telescope as only a hypothesis rather than a true description of the solar system. However, once he had what he considered proof, he wanted his description of a sun-centered universe and a moving Earth to be generally accepted.

In December 1614, a clergyman named Cassini attacked Galileo's ideas; the notion of a moving Earth was heresy, he claimed. The Inquisition ruled in 1616 that the two ideas of an immobile Sun as the center of the universe and an Earth in daily and annual motion were contrary to Church teachings. Galileo was ordered to abandon the opinion that the Earth moves through a sun-centered universe, or at least admit that such an opinion is only a hypothesis.

In the early 1600s Galileo wrote a book in which three characters discuss the role of math in understanding the natural world. After publication in 1632 the book became highly controversial. Pope Urban felt the book was a direct challenge to the authority of the Church. He ordered the book suppressed and demanded that the Inquisition charge Galileo with heresy. Consequently, Galileo was brought before the Dominican and Franciscan judges of the Inquisition in April 1633. His defense was that the motion of the Earth and stability of the Sun, although offensive to Holy Scripture, was only hypothetical. But the Inquisition decided that his punishment would be house arrest, which could be carried out at a friend's country estate. He arrived there in July 1633 at age seventy. His health was already poor, and would get worse during his remaining years.

Now he took up work begun long before, resulting in a book titled <u>Discourses and Mathematical Demonstration Concerning Two New Sciences</u>. The sciences involved the strength of materials, and the dynamic science of motion. The book laid the basis for the development of structural engineering, and discussed the science of falling objects. It explained how Galileo thought math should be used to describe physical phenomena, and how experiments were valuable to validate a mathematical hypothesis. The book was published in 1638.

Galileo's eyesight had been failing for a while; by December 1637 he was totally blind. He died in January 1642 at age seventy-nine.

The life of Galileo put the notion in many minds that there is an unavoidable conflict between science and religion. Galileo maintained that such a conflict is due to a failure of human imagination. His life showed that science had become important in human lives.

ISAAC NEWTON

Isaac Newton contributed a lot to human knowledge through his work solving certain riddles of the world. He gave the term gravity a new meaning; no longer just a serious emotion as before, but now also the invisible force that tends to push objects downward, like fruit falling from a tree. For Newton the words time, space, and motion are connected in a way that he described.

Born in 1642, as a boy he was enrolled in a school where he learned Latin, arithmetic, and theology. He hand-copied the contents of certain books that interested him. He built models of windmills, and he constructed kites that he sent aloft at night trailing lanterns, which frightened the neighbors.

In June 1661 he started to take classes at the University of Cambridge to prepare for the ministry. When an epidemic reached England, killing thousands each week, Newton returned home and spent his days reading and taking notes. He was eager to apply math to his world. He calculated the force needed to keep the moon in its orbit, and stated new principles of motion, such as if a body moves, it will continue and never stop until affected by an external cause whose force is equal to that which originally put it into motion.

At age twenty-seven Newton became a professor of math at Cambridge. But often his lectures were attended by only a few students, so he chose to lecture on light and color instead. He had sent a beam of colored light through two separate prisms, one a blue beam and one of red; when the two beams converged and were mixed, they produced white light.

Newton thought of planetary motion as maintaining a balance between a force pulling inward and another force flinging outward. He came to believe that the forces of nature – the motion of planets, comets, the moon, and tides of the sea can be expressed mathematically. In April 1686 Newton presented a book known as PRINCIPIA, describing his ideas, to the Royal Society. In it he explained the three laws of motion: *A moving body or a body at rest will remain thus unless subjected to an external force. Any change in motion is proportional to the force applied. To any action there is always an opposite and equal reaction.* In his reasoning and arguments, Newton invented a new math called calculus; but some problems he solved through geometry rather than through calculus.

In another book, Newton gathered data on the planets of our solar system, the moons of the planet Jupiter, and our own moon. He calculated the densities of the planets, and showed that the tides of our oceans are caused by the combined gravity of the sun and our moon. Although Newton explained much of the forces of nature, he admitted that there was much more work that needed to be done.

England's king named Isaac Newton Warden of the Mint in 1696, and from 1700 on, the title Master of the Mint. His job was to oversee the exchange of old coins for new ones, so he had coins subjected to rigid stands of weight and purity.

In 1703 Newton was made president of the Royal Society. He then published his early work in English rather than Latin, explaining points in words rather than through math. This book was titled <u>Opticks: a Treatise on the Reflexions, Refractions, Inflexions and Colours of Light</u>. (Spellings conformed to the usage of that time.) In it Newton stated that nature conforms only to itself; its complexity can be reduced to order. But to give meaning to observations, a theory was needed, which he, Newton, could provide. Although some claimed that Newton's theories resorted to miracles, he had based his knowledge on results of experiments; what could not be explained through experimentation he left unsolved.

Newton's death in 1727 at age 84 inspired much poetry from his countrymen. He had given the world a sense of order and security, and taken away mystery. For that he is considered one of the greatest scientists of all time.

CARL LINNAEUS

Carl Linnaeus became an inspired interpreter of nature through his writings, all of which made him a cultural figure during his lifetime and a part of the general history of science. His important contribution was the classification system he introduced for plants and animals.

He was a keen observer of his surroundings in his native country of Sweden, whether it was herbs in the meadow or insects crawling on a road. This habit began as a little boy, when his father taught him the names of flowers that grew in their neighborhood. From an early age, he explored the countryside looking for plants to collect. He admired the daisies in the fields; their golden centers seemed to reflect the sunshine that they needed to grow. Walking through the woods near his home, he would stop to inspect patches of violets. Soon he had a collection of over 600 kinds of wildflowers. Evenings he spent studying books on botany. Writing as an adult, his descriptions convey his pleasure in nature's richness, like a child among new toys.

In 1728 Carl enrolled at the University of Uppsala. In his second year there, his reputation amongst his fellow students as a lecturer grew, and administrators at Uppsala asked him to teach botany. His lectures drew crowds of three to four hundred. The beginning of a north European summer always gave Carl great happiness. Bees buzzed, birds sang, plants gave off a lovely scent. In his speeches and writings he praised nature's variety and beauty. Everything is linked together, he explained. Each creature has a part within the whole, and he considered himself the chosen interpreter of nature. Each living creature had its enemies, and thus their numbers were kept more or less constant, because such a balance had to be maintained.

Botanists in Europe in the 1700s were kept busy describing botanical specimens from distant lands. The system of descriptions of that time Linnaeus considered unsatisfactory, even chaotic. That feeling and his compulsion to arrange things into groups led him to work out, at age 23, a system to rapidly classify any plant. In 1751 he published the result of twenty years of work, which gathered his descriptions of all plant and animal forms, and explained how to name a species based on its description: first into a class, then in succession into an order, family, genus, and finally a species. A particular species was to be identified by two words: the first told the group or genus, and the second gave the species within that group. The names are all in Latin; this allows the scientific name to be the same within every language.

Botanists agreed to accept the two-word names used by Linnaeus to identify a species, and this naming system then spread to the animal world. Thus dogs became Canis familiaris and wolves became Canis lupus. The system allowed new plants and new animals to be fitted into the known framework.

As a teacher, Linnaeus had an excellent reputation; students thronged to the auditorium to hear his lectures. In Sweden he was considered his country's pride – someone who gave Swedish science a worldwide reputation. Even during his lifetime he became famous throughout the world, which gave him much satisfaction, for his goal in life was to be recognized as the chief botanist of his time. To achieve this goal he wrote reviews of his own works, calling his publications masterpieces. Easily wounded in his self-esteem, an unfriendly or critical word would lead him to complain bitterly.

At times Linnaeus suffered from depression, but he kept working. Between 1749 and 1769 he wrote 170 articles. His book Systema Naturae was revised several times; the twelfth edition, dated 1768, covered 2,300 pages divided into three volumes.

Linnaeus died January 10, 1778. A British medical student bought his collections: 14,000 pressed plants, 3,198 insects, 1,564 shells, and about 1,600 books, all moved to England. Scientists from all over still come to see and examine the work of Carl Linnaeus, the prince of botanists and the father of classification.

WILLIAM &CAROLINE HERSCHEL

William Herschel is remembered today because of his ideas concerning the structure of the universe. He excelled at building telescopes, observing and recording what the telescopes showed, and arriving at theories to explain what he saw. When his telescope showed clusters of stars, Herschel realized that tightly packed clusters were older than those with widely scattered stars, due to the effect of gravitational attraction.

The Herschels were a musical family active in music circles in their native town of Hanover in Germany, and later in England; both areas were ruled by King George II. In August 1772 William and his sister Caroline settled in Bath in western England, where William became a leading figure in the town's musical life. But he was also developing a strong interest in astronomy, which he shared with Caroline.

In 1760 William was invited to lead a small military band in northern England; this post allowed him time to also teach and compose music. In 1762, after participating in a concert in Leeds, England, city officials appointed him director of concerts.

By early 1773 William had developed a keen interest in astronomy, and over breakfast he would lecture on facts about the stars, which he had learned from a book. Only two weeks after finishing that book, he purchased a telescope, and then proceeded to learn how to construct one. Grinding and polishing a mirror was a difficult, time-consuming job. After mastering that, William then enlisted his sister Caroline and brother Alexander for his projects. But Caroline wanted only to promote her career as a singer.

Students who came to William for a music lesson were often treated to an astronomy lesson instead or in addition. He worked on polishing mirrors to use in telescopes, and built a twenty-foot reflector telescope, which allowed him to view an eclipse of the moon in July 30, 1776.

Meanwhile William had to earn a living from music, and Caroline was involved in performances of oratorios in the spring of 1777 through training sections of the chorus and singing a number of solos. After Easter the Herschels' attention returned to astronomy.

In April 1778, after a musical program that featured Caroline's singing, she was offered to come and sing in Birmingham; after a few moments' hesitation, she declined, because she felt William needed her. Her decision proved to be very important for the development of astronomy.

Up to that time astronomy had been a mathematical study of the movements of a few heavenly objects. In the late 1770s William, finding astronomy a worthy, challenging field to work in, constructed a reflector telescope that gave him an advantage over other astronomers in viewing the planets and the brightest stars.

William was invited by a friend to join the local Philosophical Society, a group of men who were interested in science. In the months that followed, William presented two papers, one on the height of mountains on the Moon, and one on a variable star known as Mira Ceti; both papers were sent to the Royal Society in London. They would make him famous in English scientific circles.

In the time he could spare from music and telescope making, William studied the brightest stars to see if they were actually two stars very close to each other. By the end of 1781 he had found 269 double stars, of which 227 were new discoveries. Other observers questioned his work, but finally came to agree with his findings.

For a long time astronomers had talked about the possibility of finding new planets in our solar system. In 1781 William reported observing an object that turned out to be a planet far away from our Sun. It was given the name Uranus. He thus doubled the size of our solar system. But his ambition reached beyond the solar system.

William experimented with various materials for his telescope. What he ended up with proved to be superior to any telescopes at England's Royal Observatory, for it showed double stars quite plainly. It was the talk of the British scientific community; he was told to show it to England's King George.

The result of all this publicity was William's decision to give up his music and concentrate on astronomy; he was to become available to England's royal family as their official astronomer, and would receive a salary from them.

William was curious about the analysis of starlight. Holding a prism between his eye and the telescope's eyepiece, he could see the light divided into different colors. In this way he examined in 1783 the spectra of six of the brightest stars; some had a lot of red and very little yellow; others contained much orange.

William's sister Caroline's career as a singer had ended, so William made her his assistant astronomer. At first she was not enthusiastic about spending cool nights outside, peering into a telescope; but after finding two new nebulas, which are distant star clusters, she changed her mind. By the end of 1783 she had found fourteen of them.

William visited some private observatories, but concluded that his telescopes were the best. Any dinner guests of the King were taken to look at the heavens through William's telescope, or else William was summoned to bring his telescope to the royal family for their viewing. Beginning in December 1782, for the next forty years the Herschels received crowds of visitors coming to view the stars through William's telescope; sometimes even King George or the Queen came. William was now mainly searching for double stars; by the end of 1784 he sent to the Royal Society his second catalogue, which contained 434 doubles. All told, his recorded observations totaled about 3,000 stars.

Caroline embraced her new career of astronomer. In the summer of 1786 she recorded having viewed 150 nebulas and a comet. In the ensuing years she found more comets; her eighth was recorded in August 1797. These discoveries made Caroline famous. In June 1802 she sent to the Royal Society a catalogue of 500 nebulas and star clusters.

Several European rulers became patrons of astronomers; William met with Napoleon, who asked him questions of astronomy, and England's King George kept up with new developments in that field. Several European kings purchased telescopes from William.

In the 1790s William published six papers dealing with the planet Saturn. His views of Uranus allowed him to calculate the mass, volume, and density of that planet.

William died in 1822.

In November 1833 Caroline, William's son John, and John's family sailed to South Africa to view southern skies. Over the next few years, they found 1,708 nebulas or star clusters, and 2,103 double stars.

In 1828 Caroline was awarded a gold medal by the British Astronomical Society for her catalogue of nebulas discovered by her brother William. In 1846 the King of Prussia awarded her a gold medal in recognition of her contribution to astronomy. She died in 1848 at age 97.

CHARLES DARWIN

Charles Darwin was an English scientist whose work completely changed the way people looked at the natural world. From something mysterious, it became something to be studied and cataloged, as scientists wondered why plant and animal species seemed so perfectly suited to each of their environments, and how they developed this suitability. Later scientists called Darwin's idea of natural selection 'survival of the fittest'.

Born in 1809 in England, from an early age Charles Darwin loved to collect things: rocks, shells from the seashore, and dead insects. He learned the name of each and made labels for them.

When he got older, Charles loved to help his brother Erasmus do chemistry experiments in a toolshed by their home. Often the two of them worked on their experiments till late at night. At age sixteen, their father sent Charles to join Erasmus at Edinburgh University to study to become a doctor. But Charles didn't like medical school; after two years his father sent him instead to Cambridge University to become a pastor. Charles found the classes there also quite dull. He became good friends with a professor with whom he took long walks. The professor taught Charles facts about many plants they found, as well as rocks and certain animals.

After graduating from Cambridge, Charles' friend, the professor, wrote him a letter inviting him to sail on a ship bound for South America, in order to study the plants and animals they would find.

Charles' father considered it a crazy undertaking, but seeing how eager Charles was, he finally consented. Charles met the ship captain in London, who took him aboard his ship.

They set sail in 1831, thus beginning a five-year trip around the world. At every stop Charles went ashore to explore. He climbed mountains and hiked in tropical rain-forests. He wrote in his journal "It is easy to specify the individual objects of admiration in these grand scenes; but it is not possible to give an adequate idea of the higher feelings of wonder, astonishment, and devotion, which fill and elevate the mind."

Where he went, he collected plants, insects, birds, and fossils. All these he sent to his professor friend in England, to be studied when he returned. Finding fossil bones delighted him especially, since they tell about life in long-ago times.

Near the end of the voyage, the ship stopped at the Galapagos Islands, off the west coast of South America. There Charles found many unusual plants and animals. But it was the birds that he shot and collected that gave him the idea that would completely change science – the principle of natural selection: that plants and animals could change over time, and the strongest of each species would survive more often.

At first Charles kept his ideas secret. Seven years after his voyage he shared his idea, and friends told him to publish. The result was his book <u>On the Origin of Species</u>, published in 1859. Copies sold out immediately.

Some people attacked Darwin's ideas, while others supported him. Darwin then published in 1871 <u>The Descent of Man</u>, in which he showed that humans could have sprung from the same ancestors as apes. He called this idea evolution. This notion offended many people. But today it is generally accepted that this and other ideas of Darwin's were correct.

GREGOR MENDEL

Gregor Mendel's mother wanted him to become a teacher, or else a priest. When he grew up in central Europe in the early 1800s, he became both, and something else besides. He became a scientist in a new field called genetics.

Genetics is the study of heredity. An example of heredity is the passing on of eye color from parent to child. Mendel's work explained how this takes place. Instead of studying human traits, he used pea plants. What, he asked himself, would happen when you cross a tall pea plant with a short one? Would they be tall, short, or somewhere in between? The resulting plant from such a process is called a hybrid.

It turned out that all of Mendel's pea plants were tall. So nature must be following some sort of rule. In 1856 Mendel began eight years of experiments with pea plants. Before he was through, he had grown over 24,000 pea plants.

First he grew two varieties of pea plants; one produced only round seeds, and the other, only wrinkled seeds. He took pollen from one kind and introduced it into the other kind of pea plant. He wrapped each pea flower with a paper bag to keep out insects that could ruin his experiments. When he opened the pea pods to look at the seeds, he found they were all round; none of the seeds were wrinkled.

The following year, Mendel planted all the round seeds. When they grew into flowers, he let them pollinate themselves, just through the work of insects or the wind. These plants produced round and wrinkled seeds in the ratio of three round to one wrinkled. Further experiments to determine seed color likewise gave a ratio of three yellow to one green. Other traits also gave the same ratio.

Then Mendel planted some of his round seeds and some wrinkled ones. The wrinkled grew into plants that made only wrinkled seeds. But the round seeds grew into plants one-third of which made only round seeds, and two-thirds became hybrid plants producing both round and wrinkled seeds in a three-to-one ratio.

Mendel thus discovered the law of nature: a one to two to one ratio (one round to two hybrid round to one wrinkled.) Round in this case is called the dominant trait, and wrinkled the recessive trait.

Mendel then turned to growing bean plants. He found that they followed the same rules as the pea plants, and later on, corn plants did also. He ended up working with seventeen kinds of plants.

In 1867 the head of Mendel's monastery died, and Mendel, at age 45, was chosen as the new abbot or chief. For three years he continued his work with plants, but then he had to give it up. Soon he also had to give up teaching.

In 1871 he ended his experiments with plants and turned to other interests, one of which was studying weather data. In 1879 he wrote a paper titled "The

Foundation for Weather Forecasting", which called for the building of many weather stations.

Mendel's work on genetics, published in 1866, was first ignored, but then rediscovered in 1900. His work proved to be important for the breeding of plants and animals.

LOUIS PASTEUR

Louis Pasteur was a noted French scientist who proved that germs produce infectious diseases; he thus increased the understanding of microbiology.

He was born in a French village in December 1822. His ancestors had been shepherds, and his father worked as a tanner, turning animal skins into leather. Hard work and patriotism were deeply ingrained in the family's values.

As a teenager Louis took up painting portraits, acting in plays, and tutoring students to earn money. He developed a love of science, and for several years studied various branches of science until, at age twenty, he passed the entrance exam to a prestigious college in Paris named Ecole Normale Superieure. The lectures he attended there and at another superb university named Sorbonne persuaded him to devote himself to the study of chemistry.

After three years of classes, time spent in a laboratory, and long hours of daily studies, an offer came to work in the Paris lab of a well-known chemist named Auguste Laurent. There Pasteur studied crystals, specifically why some bent light rays passed through them while others did not. His work launched a new field called stereochemistry – the study of how the units (molecules) of a substance are arranged spatially, and how that affects its properties.

In 1854 Pasteur was appointed professor of chemistry at the University of Lille in the fifth largest city of France. The university needed someone to help local industries that were having trouble producing alcohol from beet juice, which they needed to make vinegar, paint and perfume. Using a microscope, Pasteur saw tiny rod-shaped microbes instead of the desired round fungi called yeast. After advising factory heads what to do, Pasteur was hailed as a hero for saving the alcohol industry.

In 1859, Pasteur's oldest daughter died of typhoid fever at age nine. Deeply saddened, Pasteur resolved to learn the causes of diseases.

Many people at that time believed in spontaneous generation – that life could arise from non-living substances. Pasteur proved that disease-causing germs could not come from non-living substances. When he was asked to look into the problem of wine spoilage, which threatened the prosperity of the economy of France, he found, by using his microscope, that the spoilage could be caused by several kinds of microbes, but heating the wine for a few moments killed them. This allowed France to export wine to far-away countries, secure in the knowledge that the wine would not spoil. Likewise, heating milk prevented spoilage; this process, call pasteurization, is still used today.

Pasteur's next project was saving the French silk industry. Silk comes from the cocoon of a species of caterpillar called silkworm. After silkworms started dying in 1845, Pasteur was finally asked to help in 1865. He visited or wrote to silkworm nurseries explaining how to separate and discard sick silk moths and their eggs.

In 1877 Pasteur was asked to study anthrax, a disease that was killing cows, sheep, and horses. He confirmed that it was caused by bacteria. Weakened anthrax bacteria could prevent the disease, so Pasteur developed a vaccine that was injected into some sheep and cows, all of which remained disease-free. Requests poured in for samples of this vaccine.

Another disease that for many years caused panic and fatalities was rabies. It was caused by bites of an infected animal, usually a dog. Pasteur and his assistants, using a microscope, could not find any microbe that could be causing rabies (because the cause is a virus, not a microbe). But Pasteur's team developed a reliable vaccine: daily shots over fourteen days of increasingly stronger strains of the virus. Pasteur treated two boys, both bitten by a wild dog with rabies, and both survived. Other persons who been bitten flocked to France to see Pasteur; they were all treated and all survived.

In late 1894 Pasteur's health began to fail. He died on September 28, 1894, as a national hero. Pasteur's greatest contribution was in the field of public health. Life expectancy increased dramatically, due to an understanding of how germs are spread, and the development and availability of vaccines.

The Pasteur Institute, where scientists come to do research, opened in Paris in 1888. Its staff continues to search for ways that science can save lives.

DMITRI MENDELEEV

Dimitri Ivanovich Mendeleev was the main person to be consulted in Russia in the late nineteenth century whenever lawyers needed an expert on scientific questions, such as the inspection of cheese, poisons used in a murder, or alcohol to be measured and taxed. But nowadays he is remembered for his discovery of a law of nature and his resulting predictions that led to the formulation of the periodic table of chemical elements.

Mendeleev was born in 1834. After growing up in Siberia, his mother took him to Europe in 1850 for higher education. He thrived at St. Petersburg University, where he was encouraged to pursue his scientific interests. In April 1859 he came to Heidelberg University for further study in chemistry, and two years later he returned to St. Petersburg. Soon he had contracts to translate a German text into Russian, and to write his own organic chemistry textbook, which was well-received and gave Mendeleev a celebrated reputation. Several years later he was a chemistry professor at both St. Petersburg University and the Technological Institute; the position at St. Petersburg was the most prestigious chemistry appointment in Russia at that time.

When Mendeleev took over the post of chemistry professor at St. Petersburg University in 1867, there were 63 known elements, divided into metals and non-metals, and identified by atomic weight. This field was changing rapidly.

To help in his teaching duties, Mendeleev produced a textbook <u>Principles of Chemistry</u> in two volumes. Volume I dealt with only eight elements, leaving 55 for Volume II. For this textbook Mendeleev needed a numerical marker for each element, such as atomic weight. After listing the elements in increasing atomic weight, he observed a periodic repetition of chemical properties. First presented to beginning chemistry students, this periodic system became universally recognized after 1869, and later became a periodic law rather than simply a system. It predicted the properties of still unknown elements, based on their places according to the periodic law. Thus Mendeleev taught, as a fundamental law of science, that "all the properties of bodies are periodic functions of their atomic weights".

He died in 1907.

ERNEST RUTHERFORD

Ernest Rutherford made three important contributions to 20th century physics. He is best known for establishing the nuclear model of atoms in 1911. In addition he was able to explain the mystery of radioactivity, and he was the first to transform one element into another.

Ernest was born in 1871 in New Zealand, which was a British colony at that time. He was a good all-around scholar, with a special inclination toward a science career. In 1892 he earned a degree with honors in physics from a branch of the University of New Zealand. He then decided to do research on electricity and magnetism.

In 1895 he was awarded a scholarship in physics for study in England. His work at Cambridge University resulted in long-distance telegraphy or radio communication.

In 1898 he won a position on the faculty of McGill University in Canada. His work in the next nine years led to a better understanding of matter and energy, which included understanding radioactivity, which he called atomic disintegration. He discovered and named alpha and beta radiation, which is the spontaneous disintegration or decay of the nucleus of an atom. The alpha type emits a particle of two protons and two neutrons, while the beta is the production of high-speed electrons. These two types of radiation are usually accompanied by waves of very short wavelengths called gamma radiation (similar to light or radio waves). A full account of radioactive decay was published in 1903. The same year Rutherford was elected a

Fellow of the Royal Society in England, which was a great honor.

In May 1907 Rutherford moved to Manchester, England, to become head of the physics department at the local university. The following year he was awarded a Nobel Prize for investigations into the disintegration of the elements. Scientists noted that he had the power to inspire other scientists by his enthusiasm and interest in their research.

In 1914 Rutherford was knighted by England's King George V, thus becoming Sir Ernest Rutherford. With the outbreak of World War I, his attention moved to problems of national importance. To save British ships from attack by German submarines, he developed a method to detect and locate underwater objects by the sounds they make.

By 1917 Rutherford had shown that an atom could be broken down into smaller particles, thus changing one element into another. Whereas nature did that through the process of radioactivity, Rutherford showed that he was able to control this transformation.

In the 1930s Rutherford was part of a project to help German scientists who had been removed from their positions by the new German government. He ended up finding positions for 507 refugees. He died in October 1937 at age 66, but his work continues to inspire present-day researchers.

ALBERT EINSTEIN

Albert Einstein was one of the greatest physicists of all time. He loved to let his mind wander into thought experiments. Throughout his life he retained a childlike curiosity about nature's great mysteries, such as magnetic fields and gravity.

He was born on March 14, 1879, in the city of Ulm in Germany. After learning to play the violin as a child, he found that music helped him to think. At age seventeen he enrolled in the Swiss Federal Institute of Technology, called Zurich Polytechnic at that time. He was very good at mastering theories, but not so much at carrying out experiments. His absentmindedness became a joke among his friends.

During the time he searched for a job, he wrote papers on physics. Finally, in 1902, he joined the staff of the Swiss Patent Office, where he would remain for seven years. Those years turned out to be the most productive of his scientific life. In that job free of stress and pressure he could do what he did best: challenge any assumptions and imagine how the concepts would work in reality.

1905 became Einstein's "Miracle Year". One paper he wrote dealt with the energy properties of light and radiation. A second paper dealt with how to determine the actual size of atoms. A third explained the jittery motion of microscopic particles in liquid by using statistical analysis of random collisions. His fourth paper, based on thought experiments, would become known as the special theory of relativity and would show a relationship between mass and energy.

Einstein's thought experiments concluded that light must display a curved trajectory when it passes through a gravitational field. Einstein needed a solar total eclipse to confirm this. An eclipse in May 1919 proved that he had been correct.

Einstein came to America in 1921 and was welcomed by a motorcade traveling through cheering throngs in New York. Other cities he visited while on a grand tour of the US held parades along crowds estimated at 15,000 spectators.

Einstein was nominated for the Nobel Prize several times, beginning in 1910, but was passed over because opponents argued that relativity was more of a philosophical theory than a scientific discovery. However, he was finally awarded the 1921 Nobel in recognition of his work on the photoelectric effect, which is the emission of electrons by metals when light falls onto their surface.

In 1933 Einstein was a visiting scholar at the California Institute of Technology and decided never to return to Germany because of the political situation. Soon he was offered a position at the new Institute for Advanced Study at Princeton, New Jersey. He lived in the US for the remaining 22 years of his life.

In 1939 he sent a letter of concern to the US President, F. D. Roosevelt, about Germany's nuclear weapon industry activity, which led to the development and building of an atomic bomb. Dropping the A-bomb on Hiroshima, Japan led to an end of the war against Germany and Japan.

Einstein died in 1955.

NIELS BOHR

Niels Bohr was a scientific genius, but also a champion of peace among countries. He was born in Copenhagen, the capital of Denmark, on October 7, 1885. As a boy, his father introduced him and his younger brother to the wonders of nature and to great literature. At school he did well in math and physics, but had trouble writing essays. He completed work for a university degree in 1907, and then went on to earn advanced degrees in physics, for which he studied and wrote about the role of the electron in the structure and properties of metals. After earning his doctorate in 1911, he went to England and joined the staff at Cambridge University, where he met many scientists, and worked on mastering the English language. In March 1912 he transferred to the University of Manchester, where he worked on understanding atomic structure, which he explained in three papers published in 1913.

Bohr explained why, when a chemical substance is put into a gas flame, it produces a spectrum of light that can be separated into different colors. Each kind of atom has its own characteristic pattern of lines in its spectrum. Bohr explained that this is due to the energy levels of the electrons in the atoms of an element. Electrons are located at certain discrete energy levels or shells around the nucleus of an atom; in this way they give elements their unique chemical properties.

When back at the University of Copenhagen, Bohr asked university authorities in 1917 to establish a physics institute. This would guarantee that he would not be tempted to leave Denmark for positions abroad. The resulting Institute for Theoretical Physics opened in March 1921, and soon became the world's leading center for that branch of science. Visiting scientists appreciated that Bohr allowed them to pursue research in their own fields of interest.

In 1932 the Bohr family moved into a building called the House of Honor in Copenhagen. Its owner had specified that it should be occupied by a Dane most worthy of the honor of living there rent-free.

In 1943 a visitor told Bohr he would soon receive a set of keys from England; each key would have a tiny hole into which a secret message had been inserted. The message turned out to be from a British scientist named James Chadwick asking Bohr to come to England to work on a secret project that was aimed at helping England and the US win the war that had broken out against Germany. In December 1943 Bohr joined the project, which was already centered in the US; the resulting atomic bomb helped the US win the war.

In 1922 Bohr was awarded the Nobel Prize for physics for his work on atomic structure. His work eventually led to modern devices such as computers, medical CAT scans, and other devices that help physicians detect medical problems.

JAMES WATSON AND FRANCIS CRICK

Watson and Crick made an important contribution in the development of biology: a description, published in 1953, of the chemical substance that carries the information of life – the structure of the molecule known as DNA.

For a long time it was believed that living things were too complex to be described by the rules of science. But Watson and Crick disproved that.

James Watson was a young American scientist who graduated from the University of Chicago in 1946 and earning a doctorate in biochemistry at Indiana University, after which he went to England to do research on genes. He met Francis Crick at England's Cambridge University in 1951.

Crick was studying physics when World War II broke out. During the war he worked on underwater mines for the British, and then came to Cambridge to continue his studies. He was especially interested in the basic differences between living and non-living substances, such as proteins, viruses, and bacteria. He and Watson became friends as soon as they met, and were soon meeting for lunch almost every day. They discovered that they had many of the same ideas about biology.

One goal they both had was determining the structure of DNA, the molecule that makes up human genes. They decided to use X-rays for this – an experimental procedure called X-ray diffraction, that had been developed nearly four decades earlier. They then built a model that consisted of two DNA chains consisting of a series of smaller molecules, which were connected by bonds of a hydrogen atom; such a structure is called a double helix. To reproduce, the two chains separate, and each one then pairs with a new chain, thus making a copying mechanism for genetic material.

Soon the scientific world agreed that the structure of DNA had been satisfactorily explained.

JAMES CLERK MAXWELL

Scientists consider James Clerk Maxwell as important as Newton and Einstein in advancing human understanding of our physical world; this was accomplished through his formulation of the laws of electromagnetism into a set of mathematical equations.

Born in June of 1831 in Edinburgh, Scotland, James in his childhood exhibited a vast curiosity about the natural world, besides boundless energy, and an outstanding memory. In November 1841 he started attending an academy that became the focus of his education for the next six years; it taught the students math, physics, Latin, and Greek. At first schoolmates laughed at his country clothes and his stutter, but gradually he earned their respect.

James' father began taking him to meetings of the Royal Society of Edinburgh, which let him absorb a steady stream of scientific information. At age fourteen he wrote a paper on generalizations of the oval shape of an ellipse; his father presented that material to the Royal Society, which published it in their proceedings in April 1846.

At Edinburgh University James acquired mathematical ability. In October 1850 he transferred to Cambridge University for advanced study. In 1856 he produced two papers: one on the mixture of colors as perceived by human eyes, and one on electromagnetism. The ideas in the latter were amplified and completed in 1864, and eventually led to our modern devices, such as radio, TV, and radar.

In 1856 Maxwell joined the faculty of Scotland's University of Aberdeen, and in the early 1860s he transferred to King's College, London, becoming professor of physics and astronomy there. Besides teaching, this position allowed time for research and writing. His years in London produced publications on electromagnetic theory; one offered a math relationship between currents and magnetic fields, while another showed that light is an electromagnetic phenomenon. These papers provided new insights into the foundations of our universe, and became a turning point in the history of science. Like other great minds, Maxwell was often the first to recognize the importance of simple insights, and knowing when to apply them.

He died in November of 1877, and will be remembered as one of the few scientists who significantly altered the way in which we see the world.

RICHARD FEYNMAN

Being a physicist gives an individual the power to change the world through new discoveries or insights. Richard Feynman may be considered one of those special scientists. His way of thinking has been characterized as right-brain, perceiving patterns and emphasizing intuition, as opposed to left-brain, which seeks order and organization. According to Feynman, a scientist's work involves normal human activities, but carried out in a very exaggerated form.

He was born in 1918 in New York. In his student days, Feynman worked on a theory of the electromagnetic force which governs, among other things, the behavior of electrons that orbit the nucleus of an atom, and give atoms their chemical properties. But protons, the particles in an atomic nucleus, are subject to a force much stronger than the electromagnetic force.

In the late 1960s Feynman showed that certain observations of proton behavior could be explained by assuming protons have an internal structure of subparticles, which were named quarks. He developed a system of notation to describe and calculate subatomic reactions.

Feynman made important contributions in many areas, including optics, low-temperature physics, and electrodynamics. He avoided learning new things from books or research papers; he insisted on discovering new results himself. Research for him was a source of love, not a quest of ambition. Physics and life itself are circumscribed by intuition and inspiration, rather than by rules and customs.

In June, 1986 Feynman was a member of the Rogers Commission investigating the crash of the space shuttle Challenger. Feynman suspected the rubber O-rings forming a seal in the shuttle's solid rocket boosters (SRB) had gotten too cold in the below freezing weather at launching time. At blast-off the temperature was 28°. Instead of the rubber expanding to form a tight seal in the SRB joints, the O-rings became cold, brittle and unbendable. They failed to form a tight seal.

In a dramatic demonstration, so simple and low-key, at the hearing table, the physicist Feynman dropped a piece of O-ring rubber into a glass of ice before him (at 32° of course), where it sat for a minute. When he removed the O-ring it was brittle, with no resiliency. The small size of the demonstration contrasted sharply with the magnitude of the Challenger's explosion in the blue sky that day, an unforgettable image fresh in everybody's memory. So too the simplicity of Feynman's explanation was in contrast to the complexity of a shuttle and rocket booster coming undone in space.

This demonstration at the hearing is what Richard Feynman is remembered for, when he really should be remembered for his seven famous lectures at Cornell University on The Character of Physical Law, which can be viewed today on You Tube. In a recent interview Bill Gates had with Charlie Rose (2016), Gates stated that he watches these lectures today, replaying segments sometimes, to comtemplate more fully Feynman's simple analogies and profound explanations. Without notes, Feynman paces the stage and expands on the universe.

From 1986 on, he was weak and often depressed. In spite of several operations, cancer continued to weaken him. He survived his cancer for ten years, but it finally ended his life in February 1988 at age sixty-nine.

Feynman liked the most difficult problems – those that nobody had solved yet. But first he had to convince himself that they were worthwhile problems, and all it would take would be imagination and persistence. He wrote that he wanted to express appreciation of our wonderful world, and the physicist's way of looking at it. He seems to have succeeded in this.

www.ingramcontent.com/pod-product-compliance
Lightning Source LLC
Chambersburg PA
CBHW070331190526
45169CB00005B/1840